And God Created

Darwin

An Almighty Row

"..exploring the paradox of human evolution."

Christopher Malden

2016

*"If we think seriously about 'Creation
-ism' and 'Cognition', we encounter a
classic paradox: time couldn't have existed before the advent of human consciousness. The universe, prior to
H.sapiens, had no meaning. If our
ability to think creates an illusion, it is
still meaningless. 'Meaning' exists only
within the human mind. Outside the
human mind, time does not exist".*

Contents

In The Beginning....

There are two ways of looking at the world. One is Christian – or if you're not a Christian but Muslim... or Jewish, or Buddhist, or Hindi, etc., or belong to any one of a large number of differing religions, then a deity of some kind is deemed responsible for its creation. Myths and legends abound. Many hold as absolute truth that there is only *one* single explanation for the world's beginning and mankind's creation; theirs.

We haven't the space to tell all these histories or expound on different beliefs. Here we acknowledge them, respect them and hope not to offend by ignorance or by clumsy, or too curt a coverage of this complex and sensitive subject.

Somewhere between the two accounts—the deeply religious and the didactically scientific—may be an explanation that satisfies the criteria demanded by both sides.

One famous Englishman, who began his scientific career as a ship's naturalist, was among the first to realise that fossilised bones from rock strata millions of years old showed just how long plants and animals had existed.

Charles Darwin began to trace the evolution of species – groups of plants and animals that had similar form – seeing how they changed over time to better adapt to their own particular place in nature. Some fossils are tiny. The remains of plants and animals when pressed within rock sediments were perfectly preserved, or left an imprint of their structure. The rocks themselves tell us how long ago they lived.

LET THERE BE LIGHT...

We take the simplicity of the Christian story and simplify it even further. We have to, because the science story is also complex, so we need to simplify that, too.

Those familiar with the 'Holy Bible', the Christian book that contains the ancient account, will know the story; that the void was made light, the earth was empty, God made the seas, the land and the sky. Then God created animals and plants and also man. Taking the first man's rib, God created the first woman.

The story continues to relate how the perfect world – the Garden of Eden – was a bounteous paradise. Man and Woman could enjoy all its fruits, except those of the apple tree, for this was the seat of all knowledge. The land also contained the serpent – a force of evil.

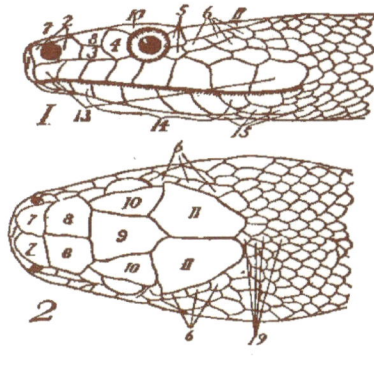

Part of the tradition of Christians is the ongoing struggle to overcome the evil with which the devil contaminated all man and woman kind, the descendents of Adam and Eve.

The serpent tempted Eve, the woman, to take the fruit and eat it, even though Adam, the man, tried to dissuade her. The cunning serpent succeeded in tempting them and after they have tasted the fruit, God is angry and banishes them from the garden.

....AND ALONG CAME SCIENCE

Of course, if God made man and woman, and they became (after a long time) scientists, then God also made science - and, in time, Charles Darwin, among others. Perhaps modern science is also just a form of temptation; like the apple - God's joke?

Alternatively, as many scientists will say, God did not create the world at all. Instead, 'God' is more likely to be a product of mankind's imagination. As humans struggled with the complexity that confronted them, a world teeming with different life forms of all shapes and sizes, they sought an explanation.

Then they found other worlds. Telescopes pioneered by Galileo in the 15th century, showed distant objects in outer space that looked very like our own. We know now there are suns in outer space with planets like ours orbiting them.

HUMAN ORIGINS....

AN ALMIGHTY ROW.

Scientists could now measure the light coming from space (and other radiation) to build a picture of how the universe perhaps came into being. Mankind's history – and that of the Earth - reveals that so much time has passed that it's pretty obvious that early beliefs didn't tell the whole story.

Given much longer time scales than provided for by the Bible, **Charles Darwin** realised that species gradually change, better to cope with changing circumstances in the natural world.

We had 'evolved' as well; our predecessors were chimpanzees and we were the branch of an older family. We still recognise the chimps as being very like humans, only they can't talk, write down sentences, invent tools. Why not ?

At most, birds and primates pick up things that are useful. All living things have been around for the same length of time, evolving from single cell organisms. Now there are ten million or more different types or species; some complex, some simple (like micro-organisms) that have been here since the beginning, *billions* of years ago. Many ancient creatures, like the shark, a species from 650 million years ago, are still with us today, virtually unchanged. Unlike other species, in just a few million years we advanced mentally, leaving forever a mode of life shaped to fit a natural niche.

Plants and animals, insects and birds all evolve, as do the smallest and the largest living creatures. Some have evolved a lot. Some have hardly changed. Perhaps these, like the shark, are the most 'successful' ones.

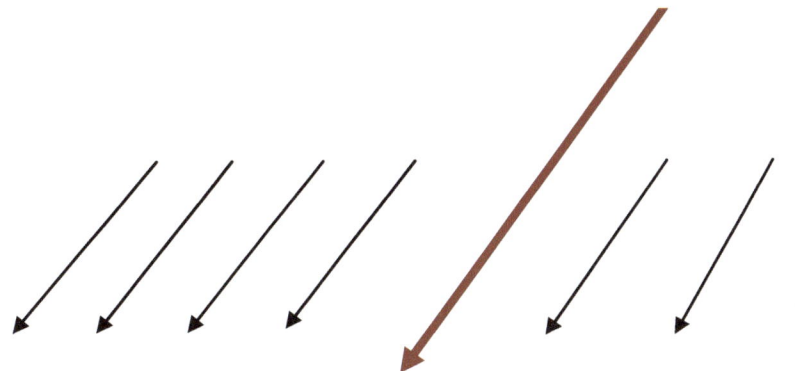

So what made *humans* change?

......It Came from Outer Space.

Radiation is an everyday fact of life. It streams down in sunlight. We are bathed in all kinds of radiation; And when fast-moving particles strike the finely balanced elements that make up the blue-print for life, changes occur, literally—out of the blue.

Over 95% of radiation—the cosmic rays that smash into the Earth's upper atmosphere are high energy protons. These originate in deep space but exactly where remains a mystery. One fact is crucial; the atmosphere is just 8 kms deep. Mutation is a fact of life, a driving force.

If successful, the 'new, improved' version has better breeding and survival chances.

All species evolve, benefitting from changes to their DNA. But evolution didn't start with Darwin—it's always been happening, from the very start, at the very start of life itself, *is* the reason there is life.

Although he knew nothing about radiation, Darwin charted the way living things evolve, pointing out that men who bred chickens, for example, seemed to be playing 'God'; creating new types—

and winning prizes at the local show—with a 'new-look' chicken, a woollier sheep, a beefier cow. Selection was already happening.

The changes to the DNA blueprint may be beneficial, subtle improvements to help exploit the resources a particular niche has to offer, perhaps to add colour to feathers or the ability to use eyes independently.

MUTATION...is the 'fine tuning' of life. Within basic cell structure is the 'blueprint' of life forms, with instructions how to construct the final individual and small errors can dramatically change improve the chances of a species to better 'fit' a physical niche. The use of the word 'fit' in evolutionary context is *not* a connotation of strength, rather it refers too the jigsaw puzzle of nature.

(Survival of the fittest' is a misconception and a cliché no thinking person adopts; it is merely an attempt to lend a false academic note to shoddy and vulgar ideas of superiority. No single species is 'superior' to any other.)

Long before humans tried - by imagining the existence of a deity - to resolve the problem of our own existence and long before Darwin, we already exploited 'natural' section, consciously selecting traits in animals that would uniquely benefit humans. But how did we come to this point?

WHAT IF THE HUMAN BRAIN...
SUFFERED A MUTATION ?

The brain of a primate is as complex as ours.
But the brain of a dolphin is even more so. Its
surface has more folds and convolutions, perhaps suggesting the level of processing power needed to decode subtle and complex sub-
marine communication. But folding, size,
complexity is not an indication of 'advanced
thinking'. Instead, *voluntary recall* and
'episodic memory' are uniquely human. A
product not of size but of plasticity.

*The gorilla ('Gorilla
sauvage' in the late 19th
C.) was once thought to
be closest to the human.*

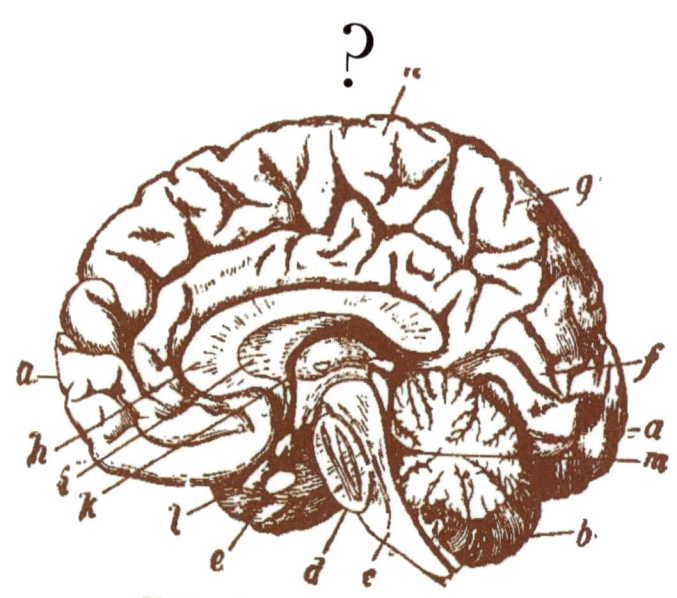

Median Section of Human Brain.

a, a, cerebrum; *b*, cerebellum; *c*, medulla oblongata; *d*, pons; *e*, pituitary body; *f*, cuneus; *g*, precuneus; *h*, corpus callosum; *i*, septum lucidum; *k*, foramen of Monro; *l*, optic nerve; *m*, fourth ventricle; *n*, paracentral lobule.

Radiation is what we call the energy streaming down from outer space. Sunlight is a form of radiation, 'particles' of energy.

Some of them are so small, so energetic they can pass right through the Earth. Most of the time they have no effect. Yet occasionally, a particle strike can upset the DNA of a living creature, plant, insect or microbe.

Orrorin
tugenensis

A.r.ramidus

Sahelanthropus tchadensis

Ardipithecus
ramidus kadat

7   6   5

Contrary to expectation, the size of the human skull bears little relation to intellectual capacity. *H. floresiensis,* a recently discovered addition to the human family, produced sophisticated tools. Yet *H floresiensis* was of tiny stature, with small skull size.

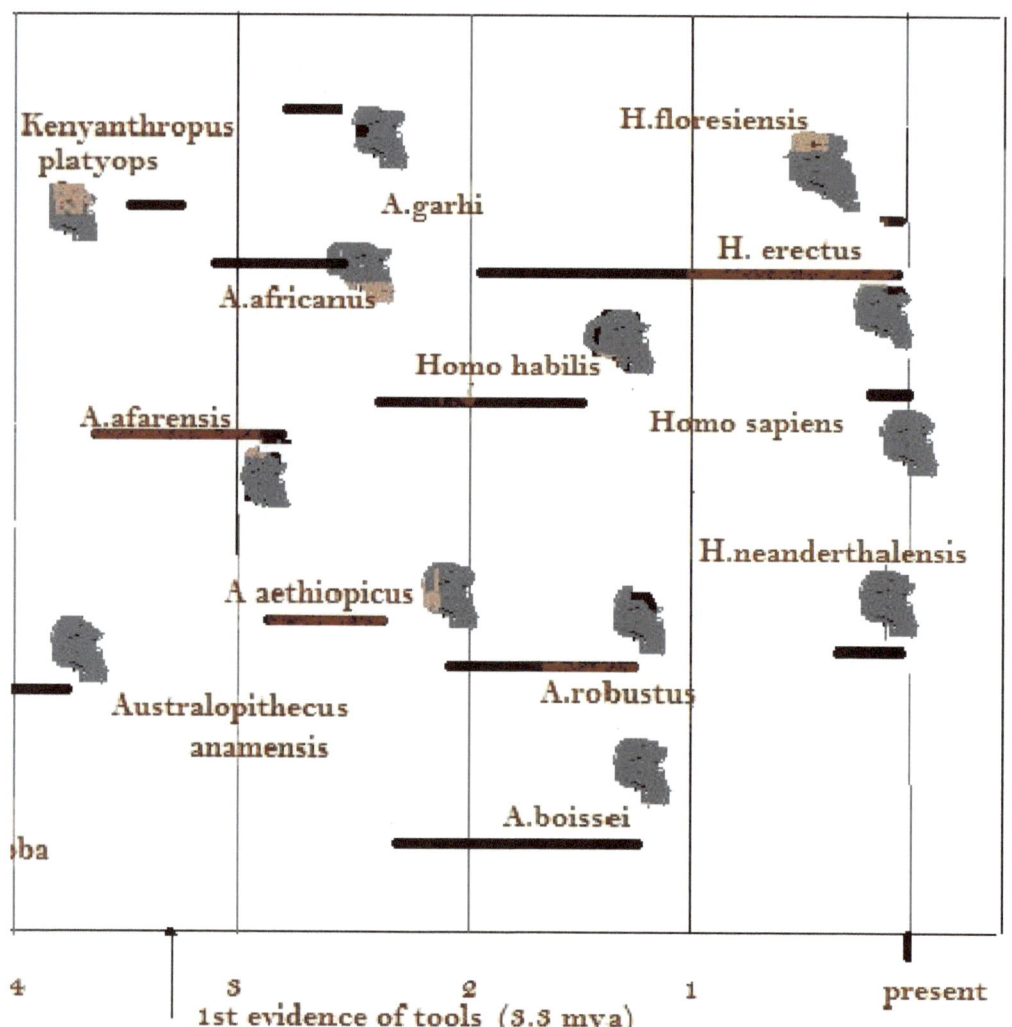

Kenyanthropus platyops

A.garhi

H.floresiensis

H. erectus

A.africanus

Homo habilis

A.afarensis

Homo sapiens

H.neanderthalensis

A aethiopicus

Australopithecus anamensis

A.robustus

A.boissei

ba

4   3   2   1   present

1st evidence of tools (3.3 mya)

Therefore a better definition of the human is that they made and make tools, skull features being irrelevant. The defining feature of the human is the way the brain is used, not size or morphology. Yet humans occupy a thin slice of recent history.

18

The 'Recall' paradigm

It seems that an abrupt change in mental ability propelled humans into a rapid evolutionary advance; instead of aeons, humans evolved in an eye-blink of time, contradicting Darwin and evolutionary theory.

Most likely, was a mutation affecting a particular function or region of the brain - it is our ability to recall events at will that defines us as human.

Uniquely, we remember past events and bring them into consciousness to aid in solving present problems. This powerful mechanism by-passes slow evolutionary processes and we build an environment to suit ourselves.

Once humans were on the move, no other animals could follow.

The evolution of 'conscious' behaviour thus presents a 'step change' in evolution.

Tool use and tool making are uniquely characteristic of the human, creating a clear distinction between us and other animals. Much has been made of the 'tools' used by other animals, but largely because of a misunderstanding; here we characterise human tools as 'manifest' tools. Those used by primates, birds and other animals as 'extemporary' tools.

Clearly, manifest tools bear the imprint of recall, evidence of *cognition* - absent in all other species. 'Extemporary' implements *cease to be useful* in the absence of an external stimulus. And are usually discarded. Cognition is inherent in man made tools, and bears a permanent imprint of the human mind..

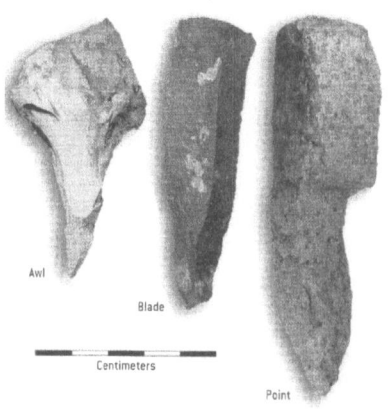

The 'manifest' tools of *H.floresiensis* (*Scientific American*)

Evidence from Eskimo and other cultures show the variety and importance of non-stone tools. Evidence of course is lost; only stone and flint being durable. This new approach concludes the first tools were degradable - fibre, skin, wood, etc.- but their undoubted presence pushes back the plausible date for human tool invention by at least a further 3 million years – perhaps more.

East=Indian Fish=traps, made of bamboo splints.

For millions of years women used, rope, fibre, thongs, leather, cane, twigs, leather, fur as 'tools' before stone blades (dated recently to 3 million years old) helped catapult the human to the forefront of evolution as a dominant predator,

H.floresiensis was a small version of the human, but produced advanced, finely-crafted, durable tools. Without doubt they made elaborate tools of leather, fibre, wood, cane, shell and bone. None of these survive. They lived long before written records but, having recall, perhaps also had ideas about a god-like 'Floresiensian' somehow responsible for their presence. More importantly they had foresight to know that, to have a chance of success, food and water had to be portable, a key identifier of the human. No other species is capable of this kind of planning or the construction the necessary kit, containers, and forward planning.

There are whale brains and spider brains, bird brains (capable also of navigating vast distances), dog and cat brains, eel brains and brains - like the killer whale, the chimp or gorilla - that apparently generate complex behaviour and forms of communication that resemble human levels of complexity, expression and subtlety. But no evidence exists of recall in other species.

Anthropocentrism is a big difficulty in determining how brains work *is the way* our brains work; of course we are influenced in our perception; because of anthropocentrism we tend to interpret the behaviour of other inhabitants of the Earth in a uniquely human way. So the anthropocentric view presents a paradox. This seems obvious, but it poses a question: are we right? Can we ever be right? Probably not. A mutation affecting the memory function of a primate ancestor is the most likely explanation for that strange creature – the human.

What happened to Darwinism?

The most difficult problem for scientists, and every other human, is that the transition from primate to human was not a smooth, gradual affair. A mutation event could explain the rapid change that separated humans from other primates. It might also be compared to an environmental catastrophe; current theory fails to account for an abrupt change so profound, so rapid, that it can be fairly said evolution *itself* evolved, with a sudden leap.

Behaviourists search in vain for signs of 'intelligence' in other species. In primates, Caledonian crows and dolphins – all we find is primate, crow or dolphin intelligence. There is no smooth evolutionary 'continuum'. linking humans to other species.

In fact 'animal intelligence' in other species may well out-perform that of the human. But human intelligence is of a different order, conscious of past, present and future.

We cannot talk to dolphins. They are 'intelligent', but with dolphin intelligence. We have at our disposal only the weak analogies that pepper the language of behavioural studies. But, in fact, we are very different.

Even our close primate cousins remain largely a mystery. Humans have advanced so rapidly that modern man falls outside any traditional schema describing species, those that evolve 'by slow degrees'

Human inventiveness has outstripped primates and all others. Recall led to cognition and inventiveness and is crucial to the definition of the human. But now- between the extremes of science and religious belief- a kind of reconciliation is possible

In Darwin's own words:

"..in the mind, it is only the comparison, with past ideas, that makes consciousness" Darwin, C. Notebook 'M',
Cambridge University Archive

While quoting Darwin liberally, endlessly and exhaustively, most evolutionary scientists are unaware of the existence of this key insight.

In Darwin's own hand, acknowledgment that recall is the precursor of human consciousness, reproduced here for the first time.

Today, *A Quietly Rational View*

If mutation caused consciousness in the human, then, from a religious perspective **a Deity might have initiated the mutation.** Whatever the religious belief, whatever the scientific argument, the paradox of human consciousness is the same: consciousness functions outside time and real events.

It turns out consciousness is wholly 'unnatural' from an evolutionary perspective. No conscious conclusions, no matter how detailed, can be consistent or 'valid', as there is no coherent system of reasoning inherent in the universe. Rational thought (the product of human consciousness) is not an echo of reason and order that somehow existed before humans. Order is an illusion that reflects, and thus neatly fits, the way we see the universe. Even human logic shows it is not 'true'. Let us put the recall theory to the test:

1.Because consciousness is the result of mutation affecting only humans, then human thought is a unique phenomenon.

2.There is no evidence of such a phenomenon exists elsewhere in the universe; there is no other sentient being 'out there'.

3   As recall is derived from a mutation, human thought (and thereby scientific reasoning), is apparently as an aberration, a random event and there is no evidence that it exists elsewhere

4   Reasoning is thus, also, an aberrant event, though it is apparently a 'benefit'; nevertheless it can also be seen as initiating the success of a runaway species - a 'pseudo species'.

5   The process of human reasoning is not as it seems; there is no universal system corresponding to an observable 'reality', to 'logic', or to a 'set of universally applicable laws'.

6 Any such 'laws' are only those that com-
   ply with the human mind-set, convenient
   'norms' that fit the 'artificial' nature of human mentality. Recall and cognition ensure
   that 'proof' (again, conveniently) is simply
   a mental construct that reflects the new
   brain architecture of the human, produc-
   ing 'self-consistency' and reinforcing
   'proof'. There is no 'echo' from the universe to verify any of its suppositions.

7 Scientific reasoning, all logical thought, is
   thus, by its own rules, fatally flawed.

8 If, for some, the sole explanation for the
   singularity of the human and their thought
   processes is a beneficent Deity, their con-
   clusion is as 'valid' as any other.

What may we CONCLUDE ?

Science is a dubious method of reasoning; a mutation (it happens all the time) conferred the ability in humans to 'step outside' the 'flow of events' (entropy) and manipulate natural objects and natural occurrences by means of a sustained and convincing illusion. We exploit to the full the massive survival advantage it confers.

Feedback of our successes, both into and from our 'mental niche, makes further expansion increasingly swift and exponentially powerful. This continues unabated.

Through this success, science has adopted the right to pronounce on all things, subverting religions in the process, and in the place of priesthoods, donning their mantle of omnipotence.

They have annexed the priestly hierarchy, have their own regalia, powerful talismanic objects eliciting reverence and awe; accelerators, space 'missions', computers.

Perhaps the notion of a final outcome – a theory of everything – is also doomed to failure, given this evidence. It is odd that a 'theory of everything' should be so very like 'belief in the almighty'. With the arrival of the human, evolution itself may be said to have 'evolved'. With recall the transient became real, the 'future' came into being, as did the 'past'. And though we know that, like time, these are not real, we still fervently 'believe'. That's how stubborn are our minds.

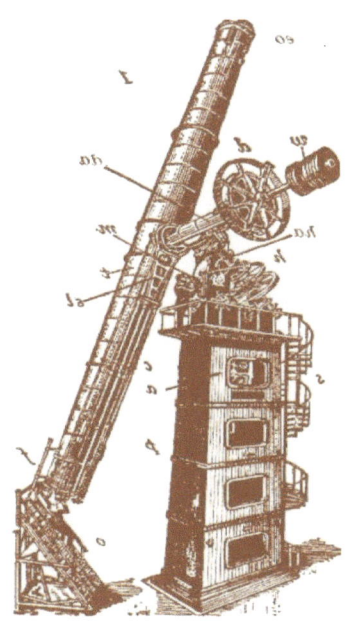

So science, after all, believes in a divine power, the human. Indeed, both are equally valid, correct and, to any individual mind, utterly convincing, even interchangeable. We cannot judge what, if anything, is 'real' . The impediment in our minds forces us to construct a mental universe which can bear no resemblance (unless we consider the human mind itself the final arbiter) to the material, transient natural world, which never repeats even an instant of 'time', nor ever duplicates a particle of matter, any event or burst of energy.

Scientists, and those of religious faith, both view the Cosmos as the embodiment of 'eternity' and that 'out there' is some kind of 'boundary', promising an end to fear or a final revelation. Beyond is peace, salvation, heaven or the Nobel Prize.

Our presentiment of death, notably the product of our special mental state involving recall 9of other deaths, is likely to be unique.

The real frontier lies in our own minds.

Ancient scriptures warn of temptation, and
that knowledge is inherently dangerous. Science may be at the point of validating this as-
sumption, as science becomes prone to errors
of complexity. Our increasing domination of
the world through the manifestation of tech-
nology may place mankind and all Earth's
other species in danger.

 Only by addressing these realities, reducing
the human impact on the biosphere, can we,
and other creatures with pedigrees far more
ancient, hope to survive and prosper.

SUMMARY

Science denies the existence of a 'higher power'. Yet science is derived from a very recent adaptation in the human brain - possibly caused by a mutation. Science now sees itself as a 'higher power', even omnipotent.

Paradoxically, strict adherence to scientific principles gives rise to this insight; it was a modified primate brain that endowed humans with cognition, thereby defining the human (albeit with a modified primate brain).

No such advanced mental behaviour is found in other species. They survive in a fixed pattern of development in a physical environment, as described by Charles Darwin. Uniquely, humans occupy a mental niche.

While denigrating religions as mere projections of patriarchal custom from early human societies, science tends to project human cognition into the universe and so hope provide a 'universal explanation', and in this way mimics deistic belief, in the form of a new, priestly caste.

Proof of Darwinian speciation is undeniable, but now a severe risk of mass extinction threatens; the biosphere faces ecological meltdown, largely because of human activity.

The 'pseudo' niche - growing exponentially in speed and size, is driven by the intense activity of a human 'pseudo species': 10 million other species are now at risk if as natural habitats disappear.

Traditional science, reasoning and beliefs are deeply flawed; 'space' is as yet uncontaminated, but the Earth itself is greatly at risk if human activity continues to accelerate.

Further reading

1859,*The Origin of Species* Darwin, C. 1871,*The Descent of Man*, Darwin, C. 1997, *The Evolution of Life on Earth*, Gould, S.J. 1990,*'The Demon Haunted World'* Sagan,C. 1994, *The Holy Bible*, *'The Making of Memory'*.Rose,S 1967, Standing, L., *Memory (*research paper). 1991,*Pollution, degradation of Nature and Species Loss.* Orr, D., 2004 *Human Molecular Genetics.* Paper *University of Chicago,* Evans, P. Et al.**2006** *'Dangerous Mind-On the Origin of Pseudo Species' (by this author).*

www.ingramcontent.com/pod-product-compliance
Lightning Source LLC
Chambersburg PA
CBHW041151180526
45159CB00002BB/783